This book belongs to

_____

Published by Sir Brody Books | Cleveland, Tennessee USA | sirbrody.com
Copyright ©D.K. Brantley, 2020 | All rights reserved

ISBN 978-1-951551-04-9

Illustrations by Adif Purnama

Building

Munching

Fishing

Spacing Out

Gardening

Patroling

Watching

Dancing

Skateboarding

Boating

Writing

Exercising

Aiming

Boxing

Rocking out

Doctoring

Motorbiking

Picking

Hanging laundry

Playing saxophone

Jumping in

Looking sharp

Juggling

Teaching

Tying

Reading

Combing

Golfing

Watching the sunset

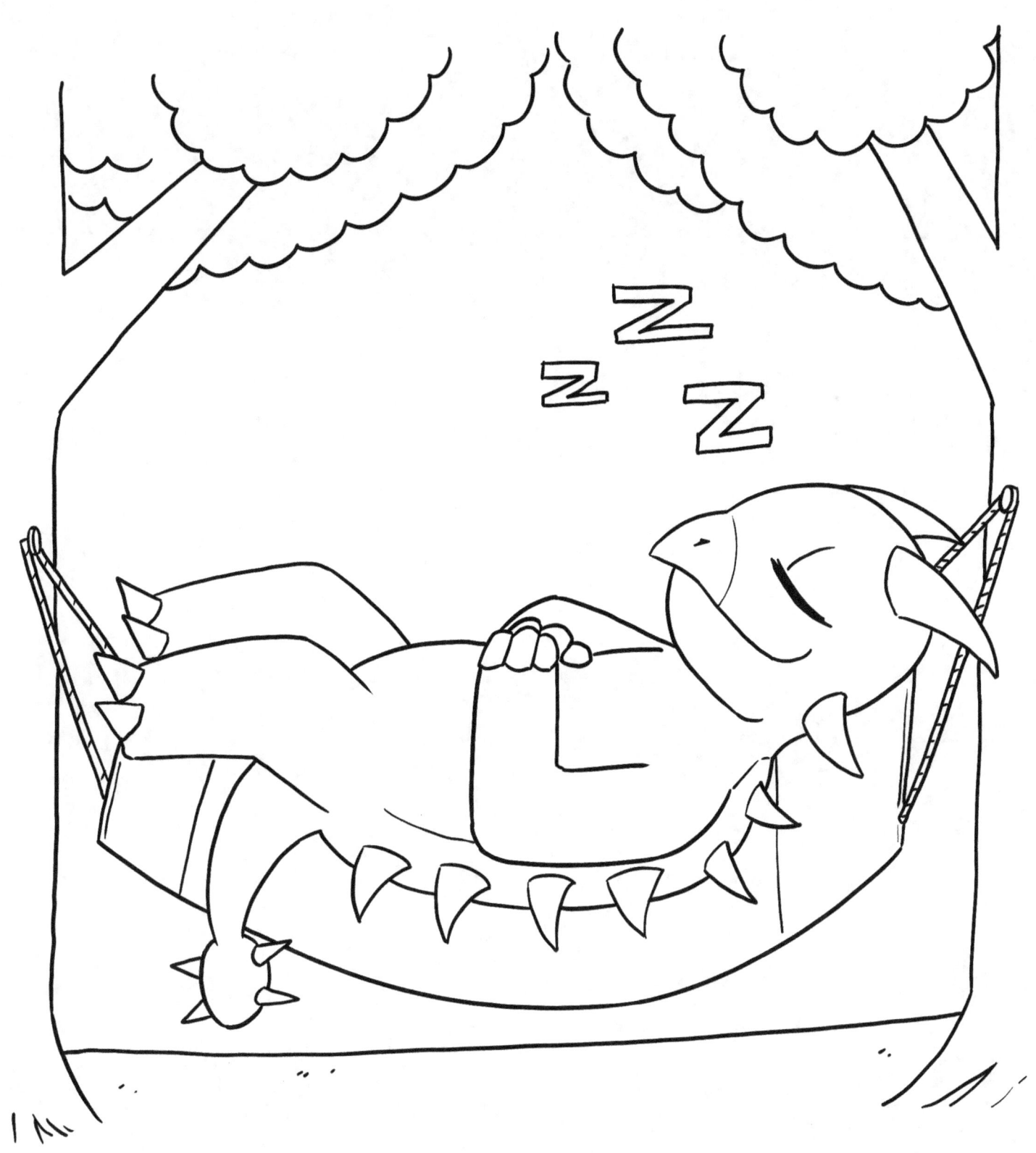

Snoozing

Bathing

Gazing

Researching

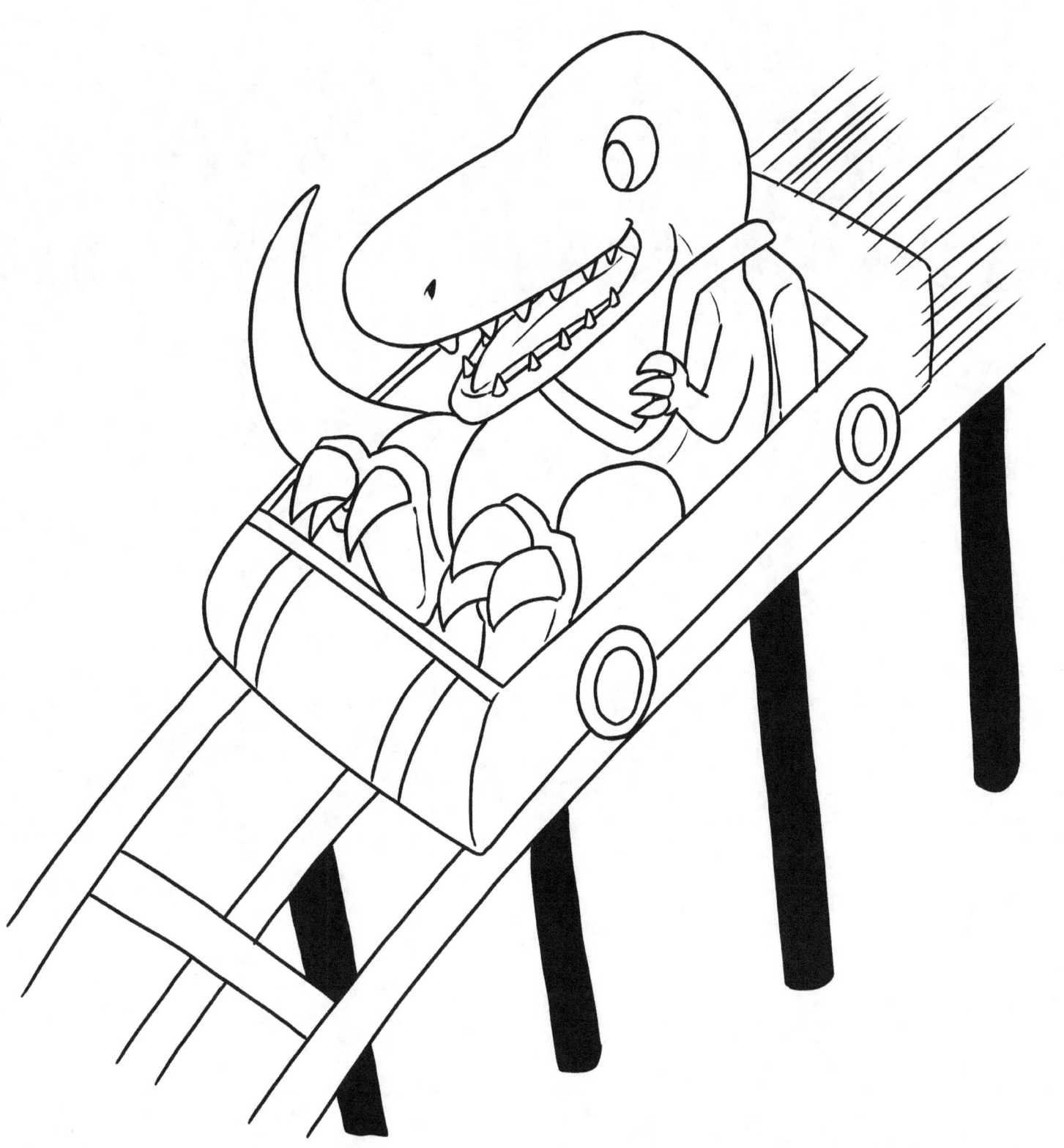

Screaming

Feeding

Building

Chatting

Judging

Listening

Firefighting

Sledding

Preparing dinner

Petting

Swinging

Making friends

Staying dry

Washing the car

Chopping wood

Texting

www.ingramcontent.com/pod-product-compliance
Lightning Source LLC
Chambersburg PA
CBHW081754100526
44592CB00015B/2422